YOUR KNOWLEDGE HAS VALUE

- We will publish your bachelor's and
  master's thesis, essays and papers

- Your own eBook and book -
  sold worldwide in all relevant shops

- Earn money with each sale

Upload your text at www.GRIN.com
and publish for free

Imprint:

Copyright © 2017 GRIN Verlag
Print and binding: Books on Demand GmbH, Norderstedt Germany
ISBN: 9783668604230

Augustine Osayande

Victim's Perception of Gully Erosion in Edo State, Nigeria

GRIN Verlag

Victim's Perception of Gully Erosion in Edo State, Nigeria

Augustine Osayande

Department of Geography and Environmental Management, University of Abuja, Nigeria

Abstract

This research is on victim's perception of gully erosion in Edo State, Nigeria. The primary objective was to evaluate how victims understand causes, effects of gully erosion and effectiveness of erosion control measures in the study area. The research used questionnaire as a tool to examined victims perception of gully erosion in the area. Out 480 questionnaire administered, 454 were returned and they were use for the analyses. Based on the findings of this study, victims of gully erosion in the area attributed causes to poor construction of culverts, deforestation and termination of drainages in sloppy topography. They agree that gully erosion in the area has resulted to losses of human lives, losses of buildings, displacement of people and losses of productive land. Victims also confirmed damages of infrastructures such as roads, bridges, buildings and altering of transportation corridors. Their responses revealed that gully erosion has resulted to decreased species richness, slowed succession and declining agricultural productivity which means less vegetation cover to soil, less return of organic matter and less biological and nutrient activity.

Key words: gully, erosion, perception, control, measures, causes, effect, effectiveness

INTRODUCTION

Soil is an important land resource and human communities depend on it's for agriculture, construction and even as a habitat. Research on soil erosion has a long scientific history and the underlying fundamentals of soil erosion processes have been investigated for many decades. But research is still ongoing and increasingly focuses on very detailed topics of gully erosion processes, causes and control measures. Mashi, Yaro and Jenkwe (2015), suggested that studies are also lacking that examine people's perception of the environmental and socio-economic consequences of gully erosion problem. According to Areola (1999), in the past people's reaction to the problem of soil erosion was to abandon the devastated area and migrate to new area. This is because they regarded soil erosion control as responsibility of government. There is need for attitudinal re-orientation of the public especially those affected by soil erosion problem. Although Areola (1999) recognized some active reaction like dumping of animal waste, ash and other household refuse materials as control measures, they have little influence as they are not used in large scale.

According to Musa, Ahmed, Muhammed and Abdul (2016), the community should be encouraged and advised to contribute their quota in addressing the problem of gully erosion through traditional means and other cultural practices such as agro-forestry system, planting of cover crops in their farms, planting trees along the streets as well as other local factors that can mitigate the gully erosion. Local communities have important roles to play in ensuring effective erosion control measures are put in place. Communities should be encouraged to form local committee to handle control of soil erosion problem. There is need for people to respond to their difficult environmental challenges by evolving highly specialized and cost effective control measures. Due to the fact that gully erosion control measures have not yielded positive result especially in study area, there is need to study the erosion control measures applied in the area and the level of community participation in application of these soil erosion control measures to reveal the most suitable gully erosion control measures and environmental conservation techniques appropriate in mitigating this problem.

STUDY AREA DESCRIPTION

Edo State is located in the South-South Zone of Nigeria. Its capital town is Benin-city. The State was created in 1991 out of the old Bendel state and its geographical coordinates are Latitudes 05° $44'$ to 07° $34'$ N and longitude 05° $04'$ and 06° $45'$ E. It has a land mass of 19,794km square and it is bordered by Kogi State to the north, Delta State to the East and South and Ondo State to the West.

The geology of the study area reveals that the entire area is underlain by sedimentary rocks. It consists of the crystalline basement rocks in the hilly and dissected zone in the north followed southwards by residual lateritic soils of the well drained dry lands at Auchi, Agbede and Afuze. Aderemi and Iyamu (2013) observed the area is underlain by sedentary rock of the Pleistocene age often referred to as the Benin formations. The sedimentary rock contains about 90 percent of sandstone and shale intercalation. It consists of over 90% sandstone, clay, shale and lignite coarse fine grained in some areas. The nature of the underlying geology contributes significantly to the origin and spread of gullies Afegbua, Uwazuruonye and Jafaru (2016). The relief of the area is mainly characterized by swamping creeks and dissected plateau such as the Esan Plateau, Orle valley and the dissected uplands of Akoko-Edo Local Government Area. According to Adediji and Felix (2013), there are six types of physical features which constitute the landscape of the area. Sandy coastal plain and alluvium clay are found in the Benin lowlands area with some hills in the east. Slopes are tilled in the southwest direction. River Osse, River Orihionmwon and lkpoba are the major drain in the area.

With the exception of River Osse that has a wide flood plain, Eseigbe and Ojeifo (2012), observed that other rivers in the area are characterised by steeply incised valleys in their upper courses and they become broad as they enter River Ethiope in Delta State. According to Eseigbe and Ojeifo (2012), the state has land mass that is relatively flat terrain in the southern part with an average height above the sea level of about 500metres except towards the northern axis where the Northern and Esan plateaus range from 183 metres at the Kukuruku Hills and 672 metres at the Somorika Hills.

The climate of the study area is humid sub-tropical indicating that it is basically within the tropical rain forest zone dominated by broadleaved trees that form dense layered stands which usually are above 50m (165 ft) in height. It is typically tropical with two major seasons- the wet and the dry seasons. Ikhile (2015), highlight that the seasons correspond to the periods of dominance of the wet tropical continental air masses and the seasonal distribution of rainfall follows the direction of the Inter-Tropical Divergence (ITD) which varies almost proportionally with distance from the coast.

The temperatures across the state is relatively high with a very narrow varies in seasonal and diurnal ranges 22-36 range with an average annual rainfall of about between 2000mm-2500mm.

The wet season comes between April and November and the dry season between December and March. According to Onakerhoraye (1995), there is a marked dry season, with duration of increases from three months in south, northwards, while the rainy period decreases inland from nine months in the south to five months in the northeast.

The vegetation zones of Edo State coincide with the political zones in the state. Edo South is in the moist rainforest, Edo Central in the dry rainforest and derived savanna and Edo North is characterized with derived savanna and southern guinea savanna. The area is also characterized by few scattered rainforests, wooded shrub lands and farmlands. Adekunle, Olagoke and Ogundare (2013), observe that the trees could be seen to be green throughout the year because they retain their leaves all through the year. This is because the temperature and precipitation are sufficiently high for continuous growth. The state is blessed with abundant natural resources. Virtually all species of hardwood can be found – high quality timber is produced from most local government areas of the State.

The state consists of eighteen Local Government Areas which include Akoko-Edo, Egor, Esan Central, Esan North-East, Esan South-East, Esan West, Etsako Central, Etsako East, Etsako West, Igueben, Ikpoba-Okha, Oredo, Orhionmwon, Ovia North-West, Ovia South-West, Owan East, Owan West and Uhunmwonde.

Major towns in the State include Benin City (the State Capital), Abudu, Ekpoma, Uromi, Auchi, and Sabongida-Ora. Generally, Edo State is traditionally known for agriculture, trade and deep historical virtues. The residents are traders and farmers whose activities are closely tied to the land. A lot of residential structures are within the gully strip, Edo State Strategic Health Development Plan (2010-2015). Due to their proximity to the gully, some of these structures have been marked as danger zones by the Edo State Ministry of Land, Housing and Survey. According to the State Strategic Health Development Plan (2010-2015) 70% of the landmass is cultivated for agricultural production as a means of livelihood with an average of about 2.345 million persons in the State directly or indirectly engage in agricultural activities. Major ethnic groups in the state are the Binis, the Esan, Ora, Etsakos, Owans, Akoko, Igarra and Afemai. The Binis occupy the southern part of the state, Esan and Ora occupy the central part and Afemai, Igarra and Akoko in the northern area. According to National Population Commission (2006) Edo State has a total population of 3,233,366. The demographic features of the area are typical of states in Southern part of Nigeria, growing rapidly with the population overstretching the weak social services. With the figure of 2006 Population and Housing Census, the state has population of 3,233,366 and an average population density for the state is about109 persons per sq. km, which is above the national average of about 96 persons per sq. km. According to the State Strategic Health Development Plan (2010-2015), the total population figures have been projected to over 3.4 million people. Most people in the area engaged in food crops cultivation. The main food crops cultivated include yam, cassava, maize and rice in the Benin lowlands and on the Esan plateau. There is also rice cultivation in the flood plains of the River Niger at Agenebode and Illushi. Tree crops such as rubber and oil palm are also cultivated in the Benin lowlands and Esan Plateau and cocoa in Owan, Etsako and Akoko Edo. The major environmental and ecological problems associated with Edo State are waste management, pollution and sanitation, forest depletion, flooding and erosion of the surface of the soil. Land degradation due to flooding and erosion ranked first and second in the objective ranking of environmental problems in the state. (Edo State Strategic Health Development Plan 2010-2015).

The magnitude of devastation as a result of flooding and erosion has resulted in loss of lives and properties, destruction of arable lands and wastage of large areas of usable lands. The State Strategic Health Development Plan also suggested need for re-afforestation, regulated construction and provision of drainage facilities in urban areas as well as attitudinal change on the part of the people.

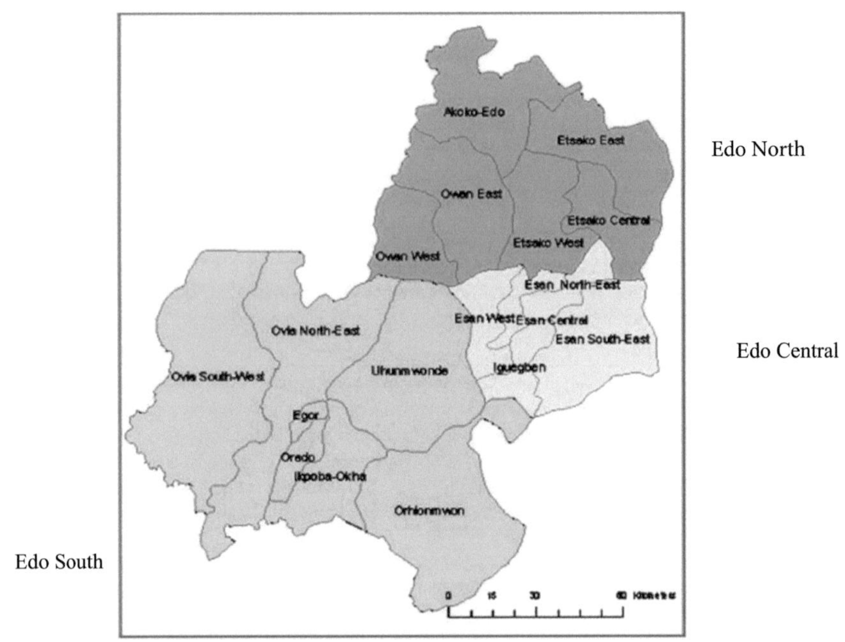

Figure 1.1: Local Government Areas in Edo State
Source: Edo State Ministry of Lands and Surveys

METHODOLOGY

The research used questionnaire to examined victims perception of gully erosion in the study area. From the 480 questionnaire administered, 454 were returned and they were use for the analyses. The questionnaire was designed to collect educational characteristic of the respondent, information on causes of gully erosion and control measures in the study area. It was also used to collect information on people's perception of effectiveness of erosion control measures applied in the area and the level of community's participation in erosion control measures in the area.

In administering the questionnaire, the purposive sampling technique was used. This system was chosen because gully erosion affects some specific areas of the state and not all the people in the study area are affected by gully erosion. The questionnaire was administered to people who are directly affected by gully erosion within 5km radius of the chosen gullies. This is to ensure that the respondents are those affected by gully erosion problem. Due to the nature of the study, considering time, cost and for convenient, a copy of the questionnaire was administered to 30 respondents at the chosen 16 gully sites making a total of 480 respondents considered in the course of this research.

RESULTS AND DISCUSSION

Age Status of Respondents in Edo North

Table 1 show that sixteen people representing 14.0% of the respondents in Edo North were within the age interval of 21-25, twelve people representing 10.5% of the respondents were within age interval of 26-30, thirty people representing 26.3% of the respondents were within age interval of 31-35, twenty four people representing 21.1% of the respondents were within age interval of 36-40, fourteen people representing 12.3%of the respondents were within age interval of 41-45, twelve people representing 12.3%of the respondents were within age interval of 46-50 and six people representing 5.3% were 51and above.

Table 1 Age Status of Respondents in Edo North

Age	Ikabigbo	Jatu Road	Egelessor	Gaz Momoh	Total	Percentage
21-25	8	-	4	4	16	14.0%
26-30	4	2	2	4	12	10.5%
31-35	6	10	6	8	30	26.3%
36-40	4	8	8	4	24	21.1%
41-54	4	4	4	2	14	12.3%
46-50	2	4	4	2	12	10.5%
51and Above	2	-	-	4	6	5.3%
Total	30	28	28	28	114	100%

Source: Field Work (2017)

Age Status of Respondents in Edo South

From Table 2 ten people representing 5.8% of the respondents in Edo South were within the age interval of 21-25, fourteen people representing 8.0% of the respondents were within age interval of 26-30, thirty two people representing 18.4% of the respondents were within age interval of 31-35, twenty four people representing 13.8% of the respondents were within age interval of 36-40, forty four people representing 25.3% of the respondents were within age interval of 41-45, twenty eight people representing 16.1% of the respondents were within age interval of 46-50 and twenty two people representing 12.6% were 51and above.

Table 2 Age Status of Respondents in Edo South

Age	Oka	Ogiso	Ikpoba Slope	Queen Eden	Evbotubu	Uniben	Total	Percentage
21-25	-	-	-	4	2	4	10	5.8%
26-30	-	4	4	4	2	-	14	8.0%
31-35	6	2	4	10	8	2	32	18.4%
36-40	4	4	4	4	6	2	24	13.8%
41-45	8	10	12	2	4	8	44	25.3%
46-50	6	4	4	2	4	8	28	16.1%
51and Above	4	2	2	4	4	6	22	12.6%
Total	28	26	30	30	30	30	174	100%

Source: Field Work (2017)

Age Status of Respondents in Edo Central

From Table 3, twenty people representing 12.1% of the respondents in Edo Central were within the age interval of 21-25, sixteen people representing 9.6% of the respondents were within age interval of 26-30, thirty two people representing 19.2% of the respondents were within age interval of 31-35, thirty six people representing 21.7% of the respondents were within age interval of 36-40, twenty four people representing 14.5% of the respondents were within age interval of 41-45, eighteen people representing 10.8% of the respondents were within age interval of 46-50 and twenty people representing 12.1% were 51and above.

Table 3 Age Status of Respondents in Edo Central

Age	AAU	Igueben	Ewohimi	Uromi	Ewu	Ibore	Total	Percentage
21-25	2	2	6	2	6	2	20	12.1%
26-30	-	4	6	-	-	6	16	9.6%
31-35	6	6	2	8	4	6	32	19.2%
36-40	10	6	6	6	4	4	36	21.7%
41-54	8	2	2	6	4	2	24	14.5%
46-50	4	2	6	-	2	4	18	10.8%
51and Above	-	4	-	6	6	4	20	12.1%
Total	30	26	28	28	26	28	166	100%

Source: Field Work (2017)

Educational Qualification of Respondents in Edo North

From Table 4, sixteen people representing 14.0% of respondents in Edo North have no formal education, twenty four people representing 21.1% has primary education. Forty four people representing 38.6% have secondary education, twenty four people also representing 21.1% have post secondary, while six people representing 5.2% fall under the category of other educational qualification.

Table 4 Educational Qualification of Respondents in Edo North

Educational Qualification	Ikabigbo	Jatu Road	Egelessor	Gaz Momoh	Total	Percentage
No formal education	2	6	4	4	16	14.0%
Primary	-	8	8	8	24	21.1%
Secondary	16	8	8	12	44	38.6%
Post Secondary	8	6	6	4	24	21.1%
Others	4	-	2	-	6	5.2%
Total	30	28	28	28	114	100%

Source: Field Work (2017)

Educational qualification of Respondents in Edo South

From Table 5, thirty two people representing 18.4% of respondents in Edo South have no formal education, forty people representing 23.0% has primary education, Fifty people representing 28.7% have secondary education; forty four people representing 25.3% have post secondary, while eight people representing 4.6% fall under the category of other educational qualification.

Table 5 Educational Qualifications of Respondents in Edo South

Educational Qualification	Oka	Ogiso	Ikpoba Slope	Queen Eden	Evbotubu	Uniben	Total	Percentage
No formal education	4	6	6	4	6	6	32	18.4%
Primary	8	8	8	8	8	-	40	23.0%
Secondary	10	6	10	10	10	4	50	28.7%
Post Secondary	6	6	6	8	6	12	44	25.3%
Others	-	-	-	-	-	8	8	4.6%
Total	28	26	30	30	30	30	174	100%

Source: Field Work (2017)

Educational qualification of Respondents in Edo Central

Table 6, shows that eight people representing 4.8% of respondents in Edo Central have no formal education, twenty eight people representing 16.9% has primary education. Sixty eight people representing 41.0% have secondary education; forty six people representing 27.7% have post secondary, while sixteen people representing 9.6% fall under the category of other educational qualification.

Table 6 Educational Qualification of Respondents in Edo Central

Educational Qualification	AAU	Igueben	Ewohimi	Uromi	Ewu	Ibore	Total	Percentage
No formal education	-	-	-	-	4	4	8	4.8%
Primary	2	4	8	6	2	6	28	16.9%
Secondary	10	12	12	14	8	12	68	41.0%
Post Secondary	12	8	6	6	8	6	46	27.7%
Others	6	2	2	2	4	-	16	9.6%
Total	30	26	28	28	26	28	166	100%

Source: Field Work (2017)

Response on Causes of Gully Erosion

For the purpose of this research, respondents were asked to identify the causes of gully erosion in the area and their response is presented in the tables below. From Table 7, Eighty Eight people representing 77.2% of respondents in Edo North attributed causes of gully erosion in the area to construction work, while twenty six people representing 22.8% attributed causes of gully erosion in the area to sloppy topography.

Table 7 Response on Causes of Gully Erosion in Edo North

Causes of gully erosion	Ikabigbo	Jatu Road	Egelessor	Gaz Momoh	Total	Percentage
Crop cultivation	-	-	-	-	-	-
Animal grazing	-	-	-	-	-	-
Construction work	30	28	26	4	88	77.2%
Bush burning	-	-	-	-	-	-
Deforestation	-	-	-	-	-	-
Sloppy topography	-	-	2	24	26	22.8%
Total	30	28	26	28	114	100%

Source: Field Work (2017)

From Table 8, Ninety six people representing 55.2% of respondents in Edo South attributed causes of gully erosion in the area to construction work, while seventy eight people representing 44.8% attributed causes of gully erosion in the area to sloppy topography.

Table 8 Response on Causes of Gully Erosion in Edo South

Causes of gully erosion	Oka	Ogiso	Ikpoba slope	Queen Eden	Evbotub u	Uniben	Tota l	Percentage
Crop cultivation	-	-	-	-	-	-	-	-
Animal grazing	-	-	-	-	-	-	-	-
Construction work	28	-	-	8	30	30	96	55.2%
Bush burning	-	-	-	-	-	-	-	-
Deforestatio n	-	-	-	-	-	-	-	-
Sloppy topography	-	26	30	22	-	-	78	44.8%
Total	28	26	30	30	30	30	174	100%

Source: Field Work (2017)

From table 9, Ninety four people representing 56.6% of respondents in Edo Central attributed causes of gully erosion in the area to construction work, twenty four people representing 14.5% attributed causes of gully erosion in the area to deforestation and forty eight people representing 28.9% of the respondent attributed causes of gully erosion to sloppy topography.

Table 9 Response on Causes of Gully Erosion in Edo Central

Causes of gully erosion	AAU	Igueben	Ewohimi	Uromi	Ewu	Ibore	Total	Percentage
Crop cultivation	-	-	-	-	-	-	-	-
Animal grazing	-	-	-	-	-	-	-	-
Construction work	30	18	10	12	4	20	94	56.6%
Bush burning	-	-	-	-	-	-	-	-
Deforestatio n	-	8	-	16	-	-	24	14.5%
Sloppy topography	-	-	18	-	22	8	48	28.9%
Total	30	26	28	28	26	28	166	100%

Source: Field Work (2017)

Effects of Gully Erosion

Gully erosion has resulted to several problems in the area. Some of the effect according to the respondent's is presented in the tables below.

From table 10, Twenty four people representing 21.1% of the respondents in Edo North have lost their land as a result of gully erosion problem; forty four people representing 38.6% of the respondents reported loss of building, while forty six people representing 40.3% of the respondents suffered cultural and economic losses in the area.

From Table 11, Twenty four people representing 13.8% of the respondents in Edo South have lost their land as a result of gully erosion problem; fifty six people representing 32.2% of the respondents reported loss of building, while ninety four people representing 54.0% of the respondents suffered cultural and economic losses in the area.

Table 12 shows that Twenty eight people representing 16.9% of the respondents in Edo Central have lost their land as a result of gully erosion problem; sixty two people representing 37.3% of the respondents reported loss of building, while seventy six people representing 45.8% of the respondents suffered cultural and economic losses in the area.

Table 10 Response on Effect of Gully Erosion in Edo North

Effect	Ikabigbo	Jatu Road	Egelessor	Gaz Momoh	Total	Percentage
Loss of land	8	10	6	-	24	21.1%
Loss of building	10	2	10	22	44	38.6%
Cultural/ Economic losses	12	16	12	6	46	40.3%
Total	30	28	28	28	114	100%

Source: Field Work (2017)

Table 11 Response on Effect of Gully Erosion in Edo South

Effect	Oka	Ogiso	Ikpoba slope	Queen Eden	Evbotubu	Uniben	Total	Percentage
Loss of land	8	4	-	2	4	6	24	13.8%
Loss of building	6	12	16	14	8	-	56	32.2%
Cultural/ Economic losses	14	10	14	14	18	24	94	54.0%
Total	28	26	30	30	30	30	174	100%

Source: Field Work (2017)

Table 12 Response on Effect of Gully Erosion in Edo Central

Effect	AAU	Igueben	Ewohimi	Uromi	Ewu	Ibore	Total	Percentage
Loss of land	14	-	6	-	-	8	28	16.9%
Loss of building	-	10	4	8	24	16	62	37.3%
Cultural/ Economic losses	16	16	18	20	2	4	76	45.8%
Total	30	26	28	28	26	28	166	100%

Source: Field Work (2017)

Gully Erosion Control Measures

The control measures used in the area include tree/crops planting, sand bagging and speed breakers, back filling and construction of channels. The response on control measures used in the area is presented in the tables below. From Table 13, thirty people representing 26.3% of respondents in Edo North chose trees and crops planting as the control measures of gully erosion, six people representing 5.3% of the respondents chose sand bagging and speed breakers, another 6 people also representing 5.3% chose back filling and seventy two people representing 63.1% chose construction of channels as the control measures of gully erosion in the area.

Table 13 Responses on Control Measures of Gully Erosion in Edo North

Control measures	Ikabigbo	Jatu Road	Egelessor	Gaz Momoh	Total	Percentage
Cultivation along contours	-	-	-	-	-	-
Trees/crops planting	24	-	-	6	30	26.3%
Sand bagging/speed breakers	4	-	2	-	6	5.3%
Control of animal grazing	-	-	-	-	-	-
Back filling	-	-	-	6	6	5.3%
Construction of Channels	2	28	26	16	72	63.1%
Total	30	28	28	28	114	100%

Source: Field Work (2017)

From Table 14, Fifty four people representing 31.0% of respondents in Edo South chose trees and crops planting as the control measures of gully erosion, forty four people representing 25.3% of the respondents chose sand bagging and speed breakers, forty two people representing 24.2% chose back filling and thirty four people representing 19.5% chose construction of channels as the control measures of gully erosion in the area.

Table 14 Responses on Control Measures of Gully Erosion in Edo South

Control measures	Oka	Ogiso	Ikpoba slope	Queen Eden	Evbotubu	Uniben	Total	Percentage
Cultivation along contours	-	-	-	-	-	-	-	-
Trees/crops planting	18	14	18	-	-	4	54	31.0%
Sand bagging/speed breakers	10	12	12	-	10	-	44	25.3%
Control of animal grazing	-	-	-	-	-	-	-	-
Back filling	-	-	-	16	-	26	42	24.2%
Construction of Channels	-	-	-	14	20	-	34	19.5%
Total	28	26	30	30	30	30	174	100%

Source: Field Work (2017)

From Table 15, Sixty eight people representing 40.9% of respondents in Edo Central chose trees and crops planting as the control measures of gully erosion, thirty people representing 18.1% of the respondents chose sand bagging and speed breakers, two people representing 1.2% chose back filling and sixty six people representing 39.8% chose construction of channels as the control measures of gully erosion in the area.

Table 15 Response on Control Measures of Gully Erosion in Edo Central

Control measures	AAU	Igueben	Ewohimi	Uromi	Ewu	Ibore	Total	Percentage
Cultivation along contours	-	-	-	-	-	-	-	
Trees/crops planting	20	14	8	18	8	-	68	40.9%
Sand bagging/speed breakers	6	2	20	2	-	-	30	18.1%
Control of animal grazing	-	-	-	-	-	-	-	-
Back filling	-	-	-	-	2	-	2	1.2%
Construction of Channels	4	10	-	8	16	28	66	39.8%
Total	30	26	28	28	26	28	166	100%

Source: Field Work (2017)

Effectiveness of Control Measures

The response on how effective is the control measures are presented in the tables below. From Table 16, twenty two people resenting 19.3% of the respondents said control measures in Edo North is not effective, sixty six people representing 57.9% said the control measures are slightly effective, twenty four people representing 21.1% said the control measures are effective, while two people representing 1.7% said the control measures are very effective.

Table 16 Response on Effectiveness of Control Measures in Edo North

	Not effective	Slightly effective	effective	Very effective	Total
Ikabigbo	6	22	2	-	30
Jatu road	8	20	-	-	28
Egelessor	8	18	2	-	28
Gaz Momoh	-	6	20	2	28
Total	22	66	24	2	114
Percentage	19.3%	57.9%	21.1%	1.7%	100%

Source: Field Work (2017)

From Table 17, fifty two people resenting 29.9% of the respondents said control measures in Edo South is not effective, sixty eight people representing 39.1% said the control measures are slightly effective and fifty four people representing 31.0% said the control measures are effective.

Table 17 Response on Effectiveness of Control Measures in Edo South

	Not effective	Slightly effective	effective	Very effective	Total
Oka	8	12	8	-	28
Ogiso	14	12	-	-	26
Ikpoba slope	14	16	-	-	30
Queen Eden	12	8	10	-	30
Evbotubu	4	8	18	-	30
Uniben	-	12	18	-	30
Total	52	68	54	-	174
Percentage	29.9%	39.1%	31.0%	-	100%

Source: Field Work (2017)

From Table 18, Forty six people resenting 27.7% of the respondents said control measures in Edo Central are not effective, ninety six people representing 57.9% said the control measures are slightly effective and twenty four people representing 14.5% said the control measures are effective.

Table 18 Response on Effectiveness of Control Measures in Edo Central

	Not effective	Slightly effective	effective	Very effective	Total
AAU	4	26	-	-	30
Igueben	2	20	4	-	26
Ewohimi	6	14	8	-	28
Uromi	8	10	10	-	28
Ewu	16	10	-	-	26
Ibore	10	16	2	-	28
Total	46	96	24	-	166
Percentage	27.7%	57.8%	14.5%	-	100%

Source: Field Work (2017)

CONCLUSIONS

Based on the findings of this study, victims of gully erosion in the area attributed causes to poor construction of culverts, deforestation and sloppy topography. This shows that victims have good awareness on causes and effects of gully erosion in the area but there is the problem of control measures implementation. They agree that gully erosion in the area has resulted to losses of human lives, losses of buildings, displacement of people and losses of productive land. Several properties whose value cannot be quantified accurately here have been destroyed and others are under treat in the area. Victims also confirmed damages of infrastructures such as roads, bridges, buildings and altering of transportation corridors. Many landlords had been sacked by the gully menace and had been turned to tenants in other parts of the town.

Victims responses revealed that gully erosion has resulted to decreased species richness, slowed succession and declining agricultural productivity which means less vegetation cover to soil, less return of organic matter and less biological and nutrient activity. A vast area of farmlands has been lost due to the menace of gully erosion while others are at their various stages of

13

destruction leading to drastic decrease in agricultural productivity and ultimately food shortage that can lead to famine.

In conclusion, gully erosion is a permanent form of soil erosion which is difficult and expensive to control. Control measures in the study area include back filling, sand bagging and speed breakers, trees and crop planting and construction of channels. These control measures were observed not to be successful in most parts of the area. Consequently, there is an urgent need to address the disaster so as to ameliorate its impact. Thus developmental agencies and policy makers should give due attention towards sustainable land management particularly soil conservation activities aimed at maintaining land productivity.

REFERENCES

Adekunle V.A.J, Olagoke A.O. And Ogundare L.F (2013), Logging Impacts in Tropical Lowland Humid Forest on Tree Species Diversity and Environmental Conservation Applied Ecology and Environmental Research 11(3): 491-511.

Aderemi, A. and Iyamu F. (2013), Risk Assessment Analysis of Accelerated Gully Erosion in Ikpoba Okha Local Government Area of Edo State. *Nigeria Environment and Natural Resources Research,* Canadian Center of Science and Education; 3(1) Pg 68-76.

Afegbua, U. K., Uwazuruonye J., and Jafaru B., (2016), Investigating the Causes and Impacts of Gully Erosion in Auchi, Nigeria. *Journal of Geography, Environment and Earth Science International* SCIENCEDOMAIN *international 4(4): 1-13, www.sciencedomain.org*

Akujieze, C.N. and Irabor E.I. (2014), Assessment of Environmental Degradation of Soil and Groundwater: A case study of waste disposal in Benin West Moat - Ekenwan gully Benin City, Edo State, Nigeria. *African Journal of Environmental Science and Technology* 8(6), pp381-390.

Areola, O. (1999), *The Good Earth: Inaugural lecture.* University of Ibadan, Nigeria. *Johnmof Printers, Ibadan.*

Edo State Government (2010), Strategic Health Development Plan (2010-2015). Edo State Ministry of Health

Eseigbe, J. O., and Ojeifo M. O., (2012), *Aspects of Gully Erosion in Benin City Edo State, Nigeria. Research on Humanities and Social Sciences* 2(7).

Ikhile, C.I., (2015), Climate change and erosion activities in Benin-Owena River Basin, South West Nigeria. *Journal of Geography and Regional Planning pp 99-110.*

Izinyon, O.C., and Ehiorobo, O.J., (2011), Measurement and Documentation for Flood and Erosion Monitoring and Control in Niger Delta Region of Nigeria. *Journal of Emerg. Trends Eng. Appl. Sci.* 3(2).

Mashi, S. A., Yaro A. and Jenkwe E. D. (2015), Causes and consequences of gully erosion: perspectives of the local people in Dangara area, Nigeria. *Environ Dev Sustain, CrossMark* 17:1431–1450

Musa D, Ahmed A. I, Muhammed U. B and Abdul H (2016), Assessment of the impacts of gully erosion on Auchi settlement, Southern Nigeria Journal of Geography and Regional Planning Vol. 9(7), pp. 128-138, July, http://www.academicjournals.org/JGRP

Onokerhoraye, A. G., (1995), Urbanization and Environment in Nigeria: Implications for Sustainable Development. *The Benin Social Science Series for Africa.* Benin City: University of Benin.

YOUR KNOWLEDGE HAS VALUE

- We will publish your bachelor's and
  master's thesis, essays and papers

- Your own eBook and book -
  sold worldwide in all relevant shops

- Earn money with each sale

Upload your text at www.GRIN.com
and publish for free